SUSTAINABILITY
Global Issues, Global Perspectives Workbook

REVISED PRELIMINARY EDITION
Edited by Astrid Cerny
New York University

Bassim Hamadeh, CEO and Publisher
Michael Simpson, Vice President of Acquisitions
Jamie Giganti, Senior Managing Editor
Jess Busch, Senior Graphic Designer
John Remington, Senior Field Acquisitions Editor
Monika Dziamka, Project Editor
Brian Fahey, Licensing Specialist

Copyright © 2016 by Cognella, Inc. All rights reserved. No part of this publication may be reprinted, reproduced, transmitted, or utilized in any form or by any electronic, mechanical, or other means, now known or hereafter invented, including photocopying, microfilming, and recording, or in any information retrieval system without the written permission of Cognella, Inc.

First published in the United States of America in 2016 by Cognella, Inc.

Trademark Notice: Product or corporate names may be trademarks or registered trademarks, and are used only for identification and explanation without intent to infringe.

Cover image copyright© 2012 by Depositphotos / 1xpert.

Printed in the United States of America

ISBN: 978-1-63487-326-0 (pbk) / 978-1-63487-327-7 (br)

www.cognella.com   800-200-3908

CONTENTS

SECTION 1: INTRODUCTION

1. An Introduction to Environmental Thought   1
BETH KINNE

SECTION 2: CHALLENGES FOR GLOBAL SUSTAINABILITY TODAY

2. Energy: Changing the Rules with Efficiency and Renewables   7
DARRIN MAGEE AND BETH KINNE

3. Human Population Explosion   15
ELISE BOWDITCH

4. Food Security and Sustainable Food Production in Africa   25
FRANKLIN GRAHAM

5. Global Climate Change and Andean Regional Impact   37
CARMEN CAPRILES

6. Our Common Good: The Oceans   43
ELISE BOWDITCH

7. The Possibility of Global Governance   53
GASTON MESKENS

SECTION 3: PARADIGM SHIFTS FOR SUSTAINABILITY

8. Education for Sustainability   61
ELISE BOWDITCH

9. Economic Schools and Different Paths to Development   69
FRANKLIN GRAHAM IV

10. Waste Management: Rethinking Garbage in a Throwaway World   75
DARRIN MAGEE

11. Ecological Landscape Practices: A sustainable model for North America   81
MIKE WILSON

SECTION I
INTRODUCTION

1. AN INTRODUCTION TO ENVIRONMENTAL THOUGHT

BETH KINNE

Critical Thinking

1. Identify three experiences in your life that have informed your idea of "environment." Explain when and where you had these experiences. Do you have positive or negative emotions about these experiences? Explain how each of these experiences influenced how you think about your relationship with the non-human environment.

2. Create an argument in favor of or against the following statement: A real "environmentalist" will prioritize the preservation of a diversity of ecosystems over the comfort, convenience and health of humans. You may need additional paper to complete this question.

3. Pretend you are Aldo Leopold. Describe the ideal relationship between humans and the land. Now pretend you are Edward Abbey and do the same.

Critical thinking and research questions

1. Choose any person or cultural tradition mentioned in this chapter. Research the environmental views of this person or tradition. Summarize what you learned about the history and origins of the environmental beliefs, values and ideas. Where and how do you see some of these views reflected in today's society?

2. Select a conservation or preservation area near where you live, attend school, work, or vacation. Read the website and educational materials for this area and summarize your findings. Identify whether the site managers are engaging in "conservation," "preservation," or a combination of the two. Support your conclusion with evidence from the documentation you find.

3. Do the people need good leaders or do leaders follow the people when it comes to innovation and change for sustainability? Choose a topic with environmental impact and argue your position.

4. Environmental protection only happens after a place has become polluted, contaminated or destroyed. Do you agree or disagree? Argue your position using real world examples.

5. Humans only act in their own self interest because they have little to no incentive to act for the common good. Using the concepts in this chapter, what is your understanding of the ideas that most shaped society in the last hundred years?

SECTION 2
CHALLENGES FOR GLOBAL SUSTAINABILITY TODAY

2. ENERGY: CHANGING THE RULES WITH EFFICIENCY AND RENEWABLES

DARRIN MAGEE AND BETH KINNE

The following questions address different topics in the energy chapter. They can be done in any order and combination to learn more about how energy production and consumption happens by state and in the United States today.

Your instructor may assign them to you as individual questions to be turned in as homework or a full lab assignment. Use the space here to take notes and compile your ideas. Write your complete answer on a separate sheet of paper.

Working with Statistical Data

ENERGY PRODUCTION AND CONVERSION:

1. Use the US Energy Information Administration website (http://www.eia.gov) to find historical data on production of two energy resources (such as coal and electricity) in the United States over the past decade. Next, plot two line graphs in Microsoft Excel or a similar program showing the trend over time for each energy resource, with time (years) on the x axis. Interpret what you see by writing a one-sentence caption for the graph.

2. Use the US Energy Information Administration website (http://www.eia.gov) to find historical data on energy consumption (use) in the United States for the following resources (see question 4) over the past decade. Next, plot a line graph in Microsoft Excel or a similar program showing the trend over time for each energy resource, with time (years) on the x axis. Interpret what you see by writing a one-sentence caption for the graph.

3. *Energy Intensity* is a measure of how much energy is used to produce one unit of a given service or product. The term is frequently used to measure how much energy a country uses to produce one unit of Gross Domestic Product (GDP). Use the US Bureau of Economic Analysis website (http://www.bea.gov) to locate historical data on GDP for the United States over the past decade. In Microsoft Excel or a similar spreadsheet program, calculate and plot energy intensity of GDP for the US over the past decade by doing the following:

    a. In Column A, list years, starting with ten years prior
    b. In Column B, list energy consumption for each of those years
    c. In Column C, list GDP for each of those years
    d. In Column D, create a calculated field to divide each cell in Column B by the adjacent cell in Column C. You may need to consult the help file for the spreadsheet program to learn about calculated fields.
    e. Plot the results of obtained in Column D against the years in Column A to show how energy intensity of GDP has changed over the past decade. What do you observe?

Working with Maps and Data Visualization

ENERGY GEOGRAPHIES:

1. Use the Internet to find the location and capacity of the following facilities nearest your hometown or campus. Capacity units are given in parentheses for each facility. Once you have found the facilities, mark their locations on Google Earth or a similar program.

    a. Wind farm (installed MW)
    b. Solar PV or solar thermal farm (installed MW)
    c. Coal-fired power plant (installed MW)
    d. Gas-fired power plant (installed MW)
    e. Nuclear power plant (installed MW)
    f. Petroleum refinery (barrels/day)
    g. Major natural gas pipeline (MCF/day or BCF/year)*
    h. *MCF = million cubic feet; BCF = billion cubic feet

   Energy conversion is a wasteful process, since some energy is downgraded to low-energy heat (thermal energy) during conversion. Some energy facilities also produce solid and liquid wastes that must be disposed.

2. Part 1: Use the Internet to find the location of the following facilities nearest your hometown or campus. Once you have found the facilities, mark their locations on Google Earth or a similar program.

    a. Nuclear waste storage facility. There are two types of nuclear waste: high-level waste, which includes spent fuel rods and weapons waste; and low-level, which includes items such as equipment and clothing used in nuclear facilities. There is currently no centralized storage facility for spent fuel in the US.

    b. Coal ash monofill. A monofill is a landfill that takes only one type of material, such as coal ash from a coal-fired power plant, or incinerator ash from a waste-to-energy garbage incinerator. These monofills are usually, but not always, located in close proximity to the power plants or incinerators they serve.

    Part 2: Evaluate the location of the energy waste facility. Is it in a good location? What makes it a good location? What do you think were the factors that determined the decision on where to site the facility? Do you see any concerns for the long term sustainability of this facility? How do you evaluate the location of the facility relative to population centers, transportation access, or protected areas?

    Part 3: (optional) If this is a site you are particularly interested in, you may be able to go to archives for a local newspaper to find out more about how the facility came to be. Or, ask a librarian for guidance. The Resource Conservation and Recovery Act (RCRA), a US federal law, which provides details on the siting of waste facilities in the United States, might also be a useful source for your research.

3. Use the US Energy Information Administration website (http://www.eia.gov) or another reliable Internet resource to find the top five countries that export crude oil to the United States. Next, consult news sources or other Internet resources to understand the geopolitical relationship between the US and those countries. Characterize the US's relationship with each exporting country as "friendly", "neutral", or "tense" based on the results of your search.

4. Locate the top five countries for solar energy and wind energy production (measured in MW of installed capacity and % of national energy consumption), using the US Energy Information Administration website (http://www.eia.gov) or another reliable Internet resource. Next, evaluate each of these countries against the energy portfolio of the United States, and consider the advantages and disadvantages that each country may have in the energy portfolio it currently relies on. Use your answers for previous questions to help you compare the US against the other countries.

Calculation Exercises

UNDERSTANDING CAPACITY AND OUTPUT:

1. Imagine that three electric power plants – one solar photovoltaic, one nuclear, and one hydroelectric – all have the same installed capacity of 1000 MW (1 GW). Assume that the capacity factors for these plants over the course of a year are 28%, 90%, and 44%, respectively. Over one year, how much electrical output will each plant produce? Express your answers in GWh or TWh. How do you interpret your results? What other information do you need in order to truly compare the three options?

2. Carnot efficiency is a way of estimating the maximum theoretical efficiency of a heat engine, including a thermal power plant. The estimate is based solely on the maximum and minimum temperatures of the steam and cooling water, respectively, and is given by the following formula:

$$E_C = \frac{T_{max} - T_{min}}{T_{max}} * 100$$, where all temperatures T are in Kelvins

Assume a coal-fired power plant has a maximum steam temperature of 580°(853 Kelvins) Celsius and a minimum cooling temperature of 10° Celsius (283 Kelvins). What is its Carnot efficiency? Could the efficiency of that power plant be doubled in theory? In practice? Explain.

Finding Sustainable Solutions

GREENING OUR ENERGY SYSTEMS:

1. Recall that the overall efficiency of a process can be no greater than the *least efficient* step in that process. Also recall that the overall efficiency of a process is the multiplicative product of all the steps in the process. The following table lists the step efficiencies for a gasoline-powered vehicle and an electric vehicle. Calculate the overall efficiency of each. What do your results suggest?

Process	Gasoline Car Step Efficiency	Electric Car Step Efficiency
Raw Fuel Production	83%	96%
Generation of Electricity	-	35%
Transmission of Electricity	-	90%
Battery	-	80%
Engine	25%	90%
Mechanical	70%	-
Transmission	70%	90%

2. Write a 500-word to 800-word letter to the editor of your hometown or university town's newspaper to make an argument for improving your town's energy efficiency. Be specific: your argument might involve residential lighting, factories, streetlights, transportation, or a similar issue. Even though YOU know that increasing end-use efficiency is more important, effective, and economical than increasing power plant efficiency, this is a difficult concept for many non-experts to grasp. You will therefore need to make a clear, concise, and convincing argument that is technically sound but easily comprehensible for readers.

3. Consider how different climates and physical geographies offer different efficiencies in the selection of energy sources. For example, in the past, oil refineries were always sited in coastal states, because it was efficient to offload the petroleum from the tanker ships and refine it nearby. Select a map of the United States or another country you know well, and fill it with energy production solutions for the future. What types of energy production facilities make sense where? Where should solar go? Are there good locations for offshore or onshore wind? Write a summary analysis of how you determined your country's energy portfolio.

4. What happens when you turn on a light switch? Create a poster, digital presentation, collage, song, or multimedia output of your choosing that answers that question in a way that is accessible for students in secondary or post-secondary school. Try to envision, illustrate, and explain all of the processes that enable this most mundane yet most revolutionary of outcomes: the lighting of an electric light bulb.

3. HUMAN POPULATION EXPLOSION

ELISE BOWDITCH

In this chapter you read about how demographers calculate changes in population over time and why exponential human population growth is a significant concern for sustainability on Earth. In the exercises below you will calculate and analyze change in human population growth for a selected group of countries. You will also explore the connection between environmental change and human migration.

Working with Statistical Data

UNDERSTANDING RAPID HUMAN POPULATION GROWTH:

Exploring Population Doubling

Choose any three countries in the world. Use a reputable source, such as the CIA World Factbook to find each country's population growth rate, and most recent gross domestic product (GDP). Complete the table with this collected data. Calculate the doubling time and the size of the doubled population.

Note: Demographers use the "Rule of 70" as a quick way to estimate the doubling time of a population given an annual growth rate. Dividing 70 by the annual growth equals the number of years it takes for the population to double. For instance, with a small annual growth rate of .01% (or .0001), population will take 70/.01 = 7000 years to double in size.

Country/Region	Current Population	Current Growth Rate	Doubling Time	Doubled Population	GDP

1. Which countries and regions have the fastest doubling time? The slowest?

2. Does there appear to be a relation between doubling time and GDP?

3. IF the total population today is just over 7 billion, what's the total percentage of the current world population for the three countries you chose?

4. What does it feel like to experience rapid population growth? Try this mental exercise. How many people live in your dormitory, your apartment building or on your block of houses? Could you live with twice as many people? Four times as many? How many times could the population double before it became physically impossible to live there? What would make this uncomfortable? What implications does this have for population size on earth?

Understanding Population Pyramids

Choose a country and look up its country's population pyramid. There are several ways to do this, for example on http://populationpyramid.net, which shows population change historically and as a future projection over a fifty year period. Draw several charts to represent each of the major shifts in your chosen country's population pyramid.

1. What stage in the demographic transition does the country appear to be in currently?

2. Does the country exhibit any unusual shapes in its "pyramid"? If so, what are some possible reasons for the unusual shape?

3. Look at pyramids for countries in the same geographic region as your chosen country. Which countries have an unusual pyramid shape for the region? What could explain the differences among countries in a region? How can you begin to explain the differences?

4. Population pyramids are snapshots in time for the year listed in the pyramid. Try to project what the pyramid will look like one generation from now. Explain your ideas and draw the shape of the future pyramid.

Discussion questions

THE IMPACT OF BIRTH CONTROL AVAILABILITY

Understanding family size:

1. Are you an only child? One of two? One of three? How many children are in your immediate family? Do you have many or a few cousins?

2. How many children do you want?

3. What advantages and/or disadvantages are there to growing up in a small versus a large family?

4. What are some reasons that couples might want large or small families?

Limiting family size:

In Agenda 21, section 3.8 j contains the following text:

> Implement, as a matter of urgency, in accordance with country-specific conditions and legal systems, measures to ensure that women and men have the same right to decide freely and responsibly on the number and spacing of their children and have access to the information, education and means, as appropriate, to enable them to exercise this right in keeping with their freedom, dignity and personally held values, taking into account ethical and cultural considerations.

1. Do you agree or disagree with this statement? Why or why not? Explain your answer with clear examples and evidence.

International governance for stable population growth:

This section of Agenda 21 also contains the following text:

> Governments should take active steps to implement programs to establish and strengthen preventive and curative health facilities, which include women-centred, women-managed, safe and effective reproductive health care and affordable, accessible services, as appropriate, for the responsible planning of family size, in keeping with freedom, dignity and personally held values, taking into account ethical and cultural considerations.

1. Do you agree or disagree with this statement? Why or why not? Explain your answer with clear examples and evidence.

2. How much support should a country's government provide for access to birth control? Should some countries provide more access than others?

3. Consider whether access to birth control is a privilege or a right. This question can be good to discuss in small groups as well as with the class. Use this space to jot down some of your ideas for a class discussion or for an assigned essay.

Collecting Data and Working with Data Analysis

MIGRATION

People migrate for a variety of reasons, such as: fleeing oppression, seeking better economic opportunities, and because environmental conditions become untenable. The drought in the United States' Midwest in the 1930s prompted migration to California. During the Potato Famine in Ireland in the late 1840s, many people faced the choice of migrating out of Ireland by boat, or starving.

Writing assignment

Identify and research one of the many potential ecological disasters being forecasted for the next 25-50 years. Several are being linked to expected climate changes, while others are linked to human activities. Consider the following criteria in writing your evaluation: why is this potential issue important, who is it likely to affect, what role does population growth play in creating the potential danger? Consider whether the environmental change in question is likely to stimulate migration or not. Which regions or countries will be affected? Evaluate the data you collect and write a descriptive analysis of the issues and the effect on human population. Use this space to collect your ideas and some of your initial data.

Discussion

Find students in class who chose a similar topic (water, drought, deforestation, invasive species, etc.), and compare the results of your research. Here are some questions that may be helpful to your group discussion.

1. Do disasters cause the same disruptions everywhere?

2. Which issues and crises in which countries would be likely to spur out migration? Which would not?

3. If a disaster such as a flood, earthquake or drought forced you to leave your home, where would you go?

4. How would you feel if any of these crises suddenly brought a sharp increase in people from other states or countries into your neighborhood?

Each group can report to the class for the general discussion.

Finding Sustainable Solutions

URBAN DENSITY AND ARCOLOGIES

Dense urban clusters are increasing in size and number, as are their ecological footprints. Traditional cities are centers of economic activity, but make heavy demands on natural resources and create massive piles of waste. Furthermore, climate change is expected to make hot climates hotter, and other climates colder, wetter, or more humid. These are environmental conditions that humans must adapt to individually, for example through protective clothing, and societally, for example through building and city design. Several experiments have been conducted over time, for example with the biosphere project in Arizona, and there are many intentional communities around the world, sometimes called ecovillages. But how can people design for ecological and social sustainability in densely populated urban clusters? Is it possible to account for rapid population growth and migration to cities?

One design solution for density is the "arcology", a term coined by architect Paolo Soleri.

Understanding arcologies

Research the concept of an *arcology*. Begin with the possible definitions, then look at existing settlements that meet the criteria, and plans for new cities. Look at the past history and evolution of the idea and even criticisms of the concept. Notice how arcologies work with extreme climates. Organize your notes and bring them to class, ready for discussion.

1. What is an arcology? What appeals to you about the idea?

2. Looking at your research results, are there any successful examples that exist? What are the elements of a successful arcology?

3. Does it even make sense to concentrate people? What are some negative consequences of concentrating people into dense clusters? What are the alternatives as human population continues to grow, and what might that mean for the global environment?

Assignment

Now that you know what an arcology is, it is time to design one. Start by choosing a climate zone. Understand its seasonal attributes, as you have learned that an arcology is adapted for and integrated into a specific climate. Then consider the population size you want to plan for and the dimensions of your new community. From here, you may consider any number of attributes, such as but not limited to building designs, transportation networks, energy production, waste management and food production. How will your arcology manage population growth?

This assignment can be completed in several parts, including blueprints, proposals, design manuals, maps, or as a research paper.

4. FOOD SECURITY AND SUSTAINABLE FOOD PRODUCTION IN AFRICA

FRANKLIN C. GRAHAM IV

The exercises for this chapter are designed to help you analyze data about three of the main topics in this chapter. The first exercise looks more closely at the exchanges discussed in the chapter, how foods flow into and out of Africa. The analysis exercise helps you understand how food security and insecurity are real time issues today, but vary across Africa. The last exercise takes a closer look at food globalization and our role as consumers of food produced by others for a global marketplace.

Your instructor may assign them to you as individual questions to be turned in as homework or a full lab assignment.

Working with Maps and Data Visualization

FOOD ORIGINS, GLOBAL TRADE AND CONSUMPTION PATTERNS

Using the internet and the five maps below, research and draw the paths a specific food item has taken throughout history, from its origins to areas of current global production. Use one map per food. Some examples of good food commodities to choose are corn, rice, wheat, coffee, tea, bananas, pineapple, peanuts, tomatoes, and cocoa, but feel free to choose any other foods that you have interests in. For clarity, label your arrows with time periods and use different colored pens or pencils to designate origin, adoption through time and current areas of production.

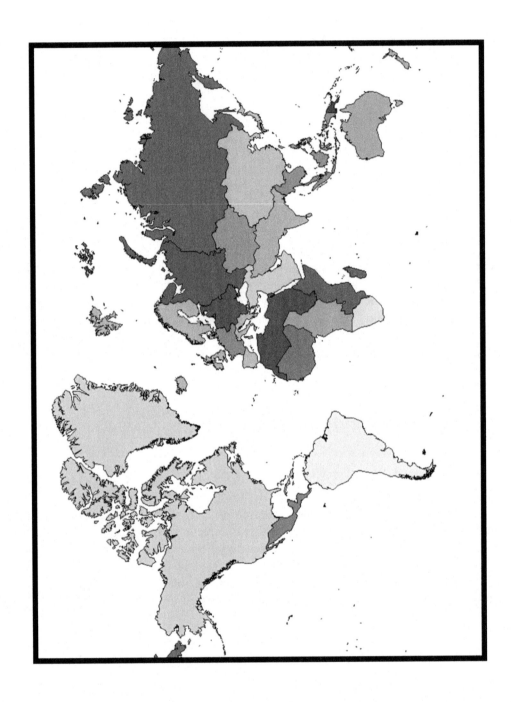

4. Food Security and Sustainable Food Production in Africa | 25

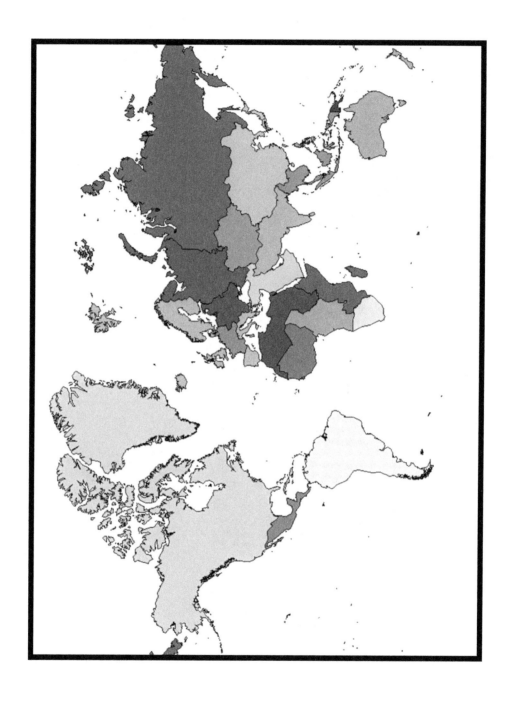

26 | Sustainability: Global Issues, Global Perspectives Workbook

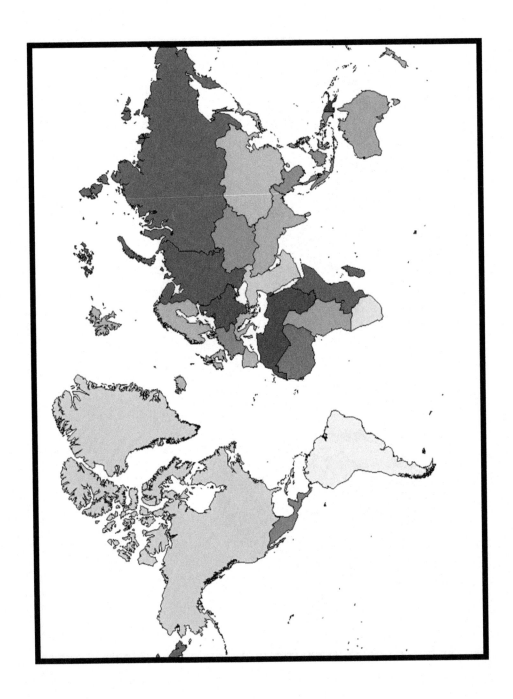

4. Food Security and Sustainable Food Production in Africa | 27

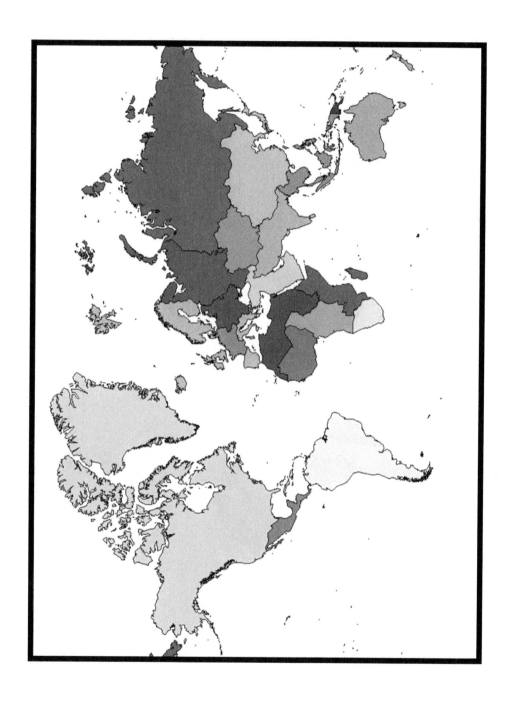

28 | Sustainability: Global Issues, Global Perspectives Workbook

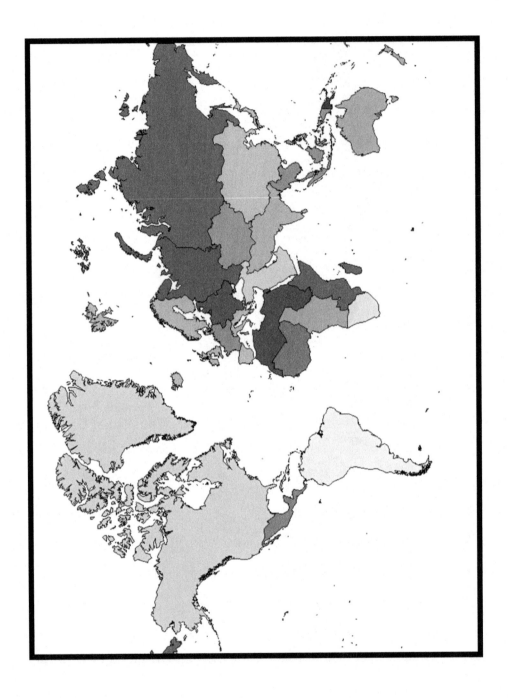

Collecting Data and Working with Data Analysis

FOOD SECURITY IN AFRICA

Below is a table displaying general statistics about African states' populations, growth rates and arable land. Based on this information, answer the following questions:

1. Where would you expect populations to be food secure?

2. Where would people be vulnerable to food insecurity?

3. Where should agricultural production be promoted? For what crops?

Assignment

Discuss your answers to these questions in a short essay or paper. Be specific and reflective about which countries you believe fit into each of the above categories. You may find it helpful to reorganize the data in this table or to do some basic calculations for the countries you are interested in.

Country	Population*	Population Growth Rate**	Arable Land**
Algeria	38,700,000	+ 1.90 %	75,501 km2
Angola	24,383,301	+ 2.78 %	33,038 km2
Benin	9,988,068	+ 2.84 %	26,029 km2
Botswana	2,024,904	+ 1.35 %	3,805 km2
Burkina Faso	17,322,796	+ 3.06 %	48,353 km2
Burundi	9,530,434	+ 3.08 %	9,124 km2
Cameroon	20,386,799	+ 2.04 %	58,868 km2
Cape Verde	518,467	+ 1.41 %	460 km2
Central African Republic	4,709,000	+ 2.14 %	19,313 km2
Chad	13,211,000	+ 1.95 %	35,258 km2
Comoros	763,952	+ 1.97 %	778 km2
Cote d'Ivoire	23,202,000	+ 2.00 %	32,531 km2
Democratic Republic of Congo	69,360,000	+ 2.54 %	64,853 km2
Djibouti	886,000	+ 2.26 %	9 km2
Egypt	87,438,500	+ 1.88 %	29,067 km2
Equatorial Guinea	1,430,000	+ 2.58 %	1,299 km2
Eritrea	6,536,000	+ 2.36 %	5,799 km2
Ethiopia	87,952,991	+ 2.90 %	112,080 km2
Gabon	1,711,000	+ 1.96 %	3,118 km2
Gambia	1,882,450	+ 2.29 %	2,788 km2
Ghana	27,043,093	+ 2.19 %	40,507 km2
Guinea	10,628,972	+ 2.64 %	10,990 km2
Guinea-Bissau	1,746,000	+ 1.95 %	2,327 km2
Kenya	41,800,000	+ 2.27 %	45,597 km2
Lesotho	2,098,000	+ 0.34 %	3,300 km2
Liberia	4,397,000	+ 2.56 %	3,304 km2
Libya	6,253,000	+ 4.85 %	18,123 km2
Madagascar	21,263,403	+ 2.65 %	29,251 km2
Malawi	15,805,239	+ 2.74 %	19,456 km2
Mali	15,768,000	+ 3.01 %	37,600 km2

Mauritania	3,545,620	+ 2.29 %	2,061 km2
Mauritius	1,261,208	+ 0.68 %	995 km2
Morocco	33,426,500	+ 1.04 %	84,797 km2
Mozambique	25,041,922	+ 2.44 %	42,576 km2
Namibia	2,113,077	+ 0.75 %	8,172 km2
Niger	17,138,707	+ 3.32 %	144,784 km2
Nigeria	178,517,000	+ 2.54 %	300,736 km2
Republic of Congo	4,559,000	+ 2.86 %	4,952 km2
Rwanda	10,996,891	+ 2.70 %	11,366 km2
Sao Tome and Prinicipe	187,356	+ 1.94 %	83 km2
Senegal	13,508,715	+ 2.51 %	24,019 km2
Seychelles	89,949	+ 0.89 %	10 km2
Sierra Leone	6,205,000	+ 2.30 %	5,694 km2
Somalia	10,806,000	+ 1.67 %	10,288 km2
South Africa	54,002,000	- 0.45 %	147,609 km2
South Sudan	11,384,393	+ 4.23 %	Undetermined
Sudan	37,289,406	+ 1.83 %	135,600 km2
Swaziland	1,106,189	+ 1.17 %	1,763 km2
Tanzania	47,421,786	+ 2.82 %	37,479 km2
Togo	6,993,000	+ 2.73 %	24,038 km2
Tunisia	10,982,754	+ 0.95 %	26,489 km2
Uganda	36,600,000	+ 3.32 %	43,077 km2
Western Sahara	586,000	+ 2.96 %	53 km2
Zambia	15,023,315	+ 2.89 %	51,777 km2
Zimbabwe	13,061,239	+ 4.38 %	40,580 km2

*Population statistics are based on three sources: (1) The Growth Rate Source: http://www.nationmaster.com/country-info/stats/People/Population-growth-rate; (2) The CIA Factbook: https://www.cia.gov/library/publications/the-world-factbook, and (3) individual Country Census.

**Population Growth Rates and Arable Land statistics come from the CIA Factbook: https://www.cia.gov/library/publications/the-world-factbook

Working with Facts and Scientific Sources

AGRICULTURE IN THE WORLD TODAY

The following questions can be useful for class discussion, or they can be answered in shorter or longer essays. The questions can be explored individually or together in greater depth.

1. What advantages do North American and European farmers have over farmers in the developing world? How do governments and private companies contribute to the disparities between Western farmers and small-scale farmers in the global South?

2. What foreseeable challenges does large-scale, industrialized agriculture face in the near future (if not already)? Create a list of these concerns and choose one from the list to explain further in your answer.

3. What similarities and differences exist between Asian and African agriculture? What obstacles does Asia face in maintaining food self-sufficiency and what role does Africa possibly have in subsidizing Asia's food needs?

4. Is it sustainable for North American, European and East Asian food markets to supply popular produce items (such as grapes, tomatoes, cherries, mangoes, cucumbers, lettuce, and others) year round? What are the social and environmental consequences should consumer demand and food retailer's capabilities continue expanding this trend? Justify your answer.

5. Is it possible for local communities to develop food production systems that are both sustainable and diverse, which provide healthy and nutritious food regimes, or is this an unrealistic proposition compared to a much needed global food system? Choose a side and justify your answer with points made in the textbook or in examples you experienced or read about elsewhere in factual sources.

Collecting Data and Working with Data Analysis

GLOBAL TRADE AND FOODS IMPORTED TO THE UNITED STATES

Global food supply chains are growing ever increasingly complex these days. For example, Coho Salmon caught off the Alaskan coast are transported to mainland China for processing and then shipped to North American and European markets where they yield a higher market price. With the exception of a few stores which insist on stocking only locally produced foods, grocery stores in the United States and Canada today carry fruits, vegetables and packaged foods produced around the world. We benefit from these imports through greater variety and selection, but how can we understand where the food in our stores, in our salad bars and on our plates comes from originally? For this assignment, you will visit a grocery store and look for foods that are labeled as to their country of origin. The parts of the assignment are best completed in order.

Assignment

1. Take notes on ten different foods in several sections of the grocery store. This can include produce, meat, fish, and packaged goods. Take notes on the identifying information for each of your ten selected foods. For example: the name of the food product, a distributing company or manufacturer, each of multiple ingredients, etc. You may find that your list of items includes those produced and packaged in the country of origin or, in some cases, both overseas and in the United States. Take thorough and complete notes at the store and organize them before class. Bring your notes and observations to class for the class discussion of the assignment.

   Note: You might be surprised to find out that some processed foods, like snacks and bottle juices come from two or more continents. Read labels carefully to evaluate the food and write your answers.

2. Using your notes from the assignment above, pair up with a partner and discuss the foods you took notes on. Where were the countries of origin? Were there ingredients from one or more countries? Does the United States grow many fruits, vegetables, grains, and produce its own meat and fish, or import more of these food products? Take notes on what you learned from the discussion here.

3. Next look at the costs of trade beyond the growing of food for people. What are the costs in transporting these foods, packaging them, marketing them? What about freshness, spoilage and waste? How does the grocery supply chain handle these, and what is the consumer's role? Take notes on your ideas and organize them in columns, as a flow chart, or some other diagram that seems to make sense to you. Identify any unknowns with a question mark and bold any conclusive results.

4. Finally, consider your food consumption options and compare them to what you read about markets and consumer choice in the chapter. What differences can you anticipate? Who has more choices today? What are the tradeoffs for modern consumers in the United States and Africa in a globalized food economy? What makes us food secure? What makes us food insecure? Follow your professor's directions for presenting your ideas for this question.

Finding Sustainable Solutions:

FOOD CONSUMPTION PATTERNS IN DEVELOPED COUNTRIES

1. Using your notes and findings from the assignment above, break up into small groups and discuss the environmental costs involved in bringing the foods each of you researched in the assignment to market. These costs can involve the farming inputs, transport, production, packaging, and food retailer operating costs. For example, these can include:

    a. how fertilizers and pesticides affect soil and water quality

    b. the creation of greenhouse gasses and pollution generated during the processing

    c. the energy expended in operating transport and processing centers

    d. the materials used for marketing and shelving the food

2. How does the carbon footprint of one food compare with other foods in your group and with the entire class? Does the price for the food item reflect the environmental impacts of production and bringing it to US markets? Why is understanding the environmental costs involved in food production growing more important today?

3. Prepare a summary report for your group on what you found out about global food production and consumption and their environmental costs.

4. Considering your own food consumption patterns, what steps does each member of your group think are doable for reducing your own negative impact? How will your behavior change the world?

5. Would food production and consumption be better off if we were not globalized today? Explain why or why not. You may wish to draw on class readings to support your position.

5. GLOBAL CLIMATE CHANGE AND ANDEAN REGIONAL IMPACTS

CARMEN CAPRILES

The following questions address different topics in the climate change chapter. They can be done in any order and combination to learn more about how complex biogeochemical processes are altered by human activities on Earth.

Your instructor may assign them to you as individual questions to be turned in as homework or a full lab assignment.

Working with Facts and Scientific Sources

CARBON IN THE ATMOSPHERE

What is your carbon footprint? How much carbon do you produce in your daily activities? Transportation is one of the main sources of carbon emissions, so let's take a closer look.

1. Calculate all the ways in which you use transportation. Are they gasoline, diesel or electric powered (trains can be any of these), whereas cars are usually gasoline. Do you fly once or several times a year? How many miles do you travel daily, weekly, monthly? Create a table that shows with columns and rows that tabulates your transportation activities.

2. Find a good source for measuring carbon and CO_2 emissions. For example: http://www.nature.org/greenliving/carboncalculator/

   Calculate your CO_2 emissions for your method of transport in question one. You can also do this for your household energy consumption and other topics in a separate table or for your own notes to understand the carbon footprint concept more fully.

The trouble with carbon is that we are emitting it into the atmosphere in great quantities at many scales. Leaving the scale of the individual, consider the footprint of such things as a farm, a small town, a large city, an airline, a trucking company, the strawberry industry, the beef industry or a sports team.

1. Choose a carbon emitter from this list or pick your own and try to figure out its annual carbon emissions and its carbon footprint.

2. First you will need to collect your data, and to do that, you'll need to decide what you are collecting. So break down the emitter you chose into the parts you think are relevant to study. Perhaps write a list, or take informal notes.

3. With the help of a librarian or on your own, find and collect data about emissions, pollution, and waste produced by your emitter. Ideally, you will be working with published numbers from a reliable source, such as a statistical yearbook or an annual report.

4. Collect your quantitative and qualitative findings and organize them into a report about your research. Write your own summary findings.

ICE MELTING

The melting of high altitude glaciers, the thawing of the Greenland ice sheet, as well as changes in the Arctic and Antarctic ice are results that have been scientifically observed over longer periods of time. The data are comprised of systematic efforts to understand ice and polar conditions, but also to record changes over time in environmental indicators.

Locate the Arctic report card at the NOAA website.

1. What are the significant environmental indicators that NOAA monitors? Make a list of five to ten indicators and summarize what each indicator tells us and helps us to understand.

2. Take a blank, unlined piece of paper and use a pencil to draw a mind map. Start with the indicators you just wrote about above and put them into your map as keywords. Draw lines and arrows to show how they are connected. Using your textbook chapter on climate change, and the NOAA website, add more keywords to expand your map and add details.

3. Bring your mind map to class and be prepared to discuss how you conceptualized your map with a small group. You may wish to take notes here to have your ideas handy in class.

Finding Sustainable Solutions

SHARING THE BURDEN OF REDUCING CARBON EMISSIONS

In the chapter you learned how greenhouse gasses become trapped in the atmosphere and how global warming is affecting different parts of the world. You learned how climate change is expected to affect regions disproportionately and regardless of which countries contributed more or less to the current situation.

Global movements have arisen as citizens become motivated to act, typically by one of two factors. First, their governments are not acting quickly enough or at all, and second, people feel a sense of solidarity with other countries and parts of the world than the one they happen to live in.

One such movement is the People's Climate March, which took place in New York in September 2014. Another is the People's Sustainability Treaties, which were written at Rio +20 in Rio Di Janeiro in June 2012.

In both of these examples, individuals that had strong values about the importance of finding solutions came together, sometimes representing themselves, but more often representing communities and small organizations.

1. Is there strength in numbers? Are there right and wrong sides to be on? Can such international events be effective at creating change? How?

2a. Find a movement, rally, protest, exhibition or other event from the last five years in any part of the world, whose subject was global warming or climate change.

2b. Read about it to understand its purpose, how and when it happened and the results reported about it.

2c. Analyze the event based on whether you think it was effective or not. Discuss your analysis in a short paper.

Recommended Books

Tim Flannery (2001) *The Weather Makers: How Man Is Changing the Climate and What It Means for Life on Earth*

James Hansen (2006) *Storms of My Grandchildren: The Truth about the Coming Climate Catastrophe and Our Last Chance to Save Humanity*

Naomi Klein (2014) *This Changes Everything: Capitalism vs. The Climate*

Bill McKibben (2007) *Fight Global Warming Now: The Handbook for Taking Action in Your Community*

E. Kirsten Peters (2012) *The Whole Story of Climate: What Science Reveals About the Nature of Endless Change*

Recommended Reports

IPCC Working Group (2013) Climate Change 2013: The Physical Science Basis http://www.climatechange2013.org/images/report/WG1AR5_SPM_FINAL.pdf

World Bank (2013) Turn Down the Heat, Why a 4°C Warmer World Must be Avoided Climate Extremes, Regional Impacts, and the Case for Resilience https://openknowledge.worldbank.org/bitstream/handle/10986/20595/9781464804373.pdf?seuence=3

6. OUR COMMON GOOD: THE OCEANS

ELISE BOWDITCH

Below are exercises which deepen and expand a few of the topics you read about in the oceans chapter.

Your instructor may assign them to you as individual questions to be turned in as homework or as a full lab assignment. The topics can be studied sequentially or out of order, and the questions can be used to write different types of scientific writing, from a short data analysis to a longer paper.

Collecting Data and Working with Data Analysis

OCEAN CURRENTS

Climate

For a simple but animated discussion of ocean currents, view "The Gulf Stream & Climate Change" at https://www.youtube.com/watch?v=UuGrBhK2c7U (5 minutes). Then, for a more detailed understanding, read the National Ocean and Atmospheric Agency's discussion of ocean currents (pages 1-3) at http://oceanservice.noaa.gov/education/kits/currents/06conveyor2.html

1. Write your own definitions to help you understand these key terms:

    a. Surface ocean currents

    b. Deep ocean currents

    c. Coriolis effect

    d. Density

    e. Thermohaline circulation

2. For discussion or as a writing assignment:
   NOAA says that changes in ocean saline content and temperature "…could slow or even stop the conveyor belt, which could result in potentially drastic temperature changes in Europe."

   Looking at the map(s) of global ocean circulation on the NOAA website, would Europe be the only place that suffers if ocean currents change? What other regions of the earth would experience rapid change if the currents that have influenced their climate for millennia were to shift course or cease? Choose a region, and explain what might happen in this case.

Trash

View another animation of ocean currents, "Perpetual Ocean", at http://www.livescience.com/19662-animation-reveals-ocean-currents.html

1. Explain what is different about this animation compared to the previous two video selections

2. Based on what you see in this animation, where might you expect trash and plastic debris to concentrate? Why?

3. Compare this animation to the map and discussion of "Trash Islands" at http://ocean.si.edu/ocean-news/ocean-trash-plaguing-our-sea

4. Read Richard Matthew's article "Plastic Waste in Our Oceans: Problems and Solutions" at http://globalwarmingisreal.com/2014/04/10/ocean-garbage-problems-solutions/ and prepare some scientific notes on what you learned from reading the article.

5. In the article, the author mentions cleanup methods and plans that have been proposed at one time or another. Choose one of these plans, and research the proposal in more depth. Could it work? Why? Where would it fail? Try to come up with your idea of a tool, method or law that could reduce the problem of ocean plastics. Summarize your idea in writing.

Finding Sustainable Solutions

PLASTICS IN THE OCEANS

Scientists do not always agree about what the solutions to our environmental problems. Scientists argue amongst themselves and propose new ideas that may discount or oppose other ideas. This is considered healthy scientific debate. Review one such forum at: http://inhabitat.com/the-fallacy-of-cleaning-the-gyres-of-plastic-with-a-floating-ocean-cleanup-array/

1. For discussion or as a writing assignment:
   Do you agree or disagree with Stiv Wilson's assessment that "the vast majority of the scientific and advocacy community believe it's a fool's errand – the ocean is big, the plastic harvested is near worthless, and sea life would be harmed. The solutions start on land"?

2. Who and what does Stiv Wilson not agree with? What is his scientific answer to the problem of ocean trash? How can you assess the scientific credibility of Stiv Wilson's ideas and the article itself?

Critical Thinking:

THE FISHERIES

Trophic Levels and Cascades

Is there a marine environment you are interested in, even if you have never been there? Or one that you grew up around, know and care about? There is much to learn about what is happening beneath the water surface.

1. Choose a marine ecology that appeals to you from any region in the world that you want to know more about.

    Some examples of a marine ecology:

    a. Iceland
    b. Irish Sea
    c. The Galapagos Islands
    d. The Gulf of Mexico
    e. Lake Baikal
    f. Lake Tanganyika
    g. North Sea
    h. The Caribbean Sea
    i. The South China Sea
    j. West African coast
    k. Tyrrehenian Sea
    l. Chesapeake Bay, MD
    m. Willapa Bay, WA

2. Find out the current state of the fisheries in that area. What are the dominant fish species? What are the popular species? How are they marketed?

TIP: See if you can discover any historical material on fisheries in this area for what existed a century or more ago. There is no backwards time limit, as some places may have information going back centuries while others are limited to a few decades. Use a good library database for this exercise, do not rely on the internet only.

Some possible questions to answer:
1. Are the species the same or different than what is currently fished? If a species has been fished for a long time, can you find a prognosis for the future of the fish population?

Measuring Fish

You read about the problem of overfishing in the chapter. In order to avoid overfishing, scientists and the fishing industries try to estimate the number of fish they can take out of the ocean. There are three main methods for estimating the annual allowable catch. One is **constant catch**. Regardless of how many fish might be there, allow some number of tons to be caught each year. Another is **constant proportion**. Under this method, estimate the total biomass (total weight of fish), and allow some percentage to be caught each year. The last is **constant escapement**. Estimate the biomass necessary for the fish species' survival, and allow any catch above that level. Theoretically, this allows the Optimal Sustainable Yield.

Questions for discussion or as a writing assignment:
1. What problems do you see with each method for a given fish species, and for any varieties of fish species that might be found together in a section of ocean fishing grounds?

2. What problems might exist for fishers using each method?

3. What are the theoretical differences between *Maximum* Sustainable Yield and *Optimal* Sustainable Yield?

Protecting marine resources

Some sources of articles about marine parks are at: http://theconversation.com/us/topics/marine-parks and   http://www.wri.org/resource/marine-protected-areas-world

A comprehensive list of data about marine parks is at http://www.protectedplanet.net (search for "marine" in the search box, and use the menu on the left in the results to narrow your choices.)

4. Using the above, or your own resource if you have one, choose a marine park (or reserve, or sanctuary) in a place that interests you. Research the park in more depth, and answer the following questions:

   How long has it been in existence? What prompted people to create a marine park (or reserve or sanctuary) at this place? What human activities are allowed within it or prohibited? How big is it? What species and ecosystems does it protect? What are the challenges and threats to its continued existence? What else would you want people to know about this place?

Finding Sustainable Solutions

WHAT TO EAT?

1. List three to five of your favorite types of fish and seafood. If you don't eat marine foods, ask someone you know outside of class such as a friend or family member and use that list.

2. Many organizations keep track of species that are commercially fished. They publish guides for eating so that consumers can make informed decisions and avoid seafood that is endangered. Some supermarkets have their own, and respected fisheries related organizations such as the Monterrey Aquarium.

    A good place to start is this handy guide that provides an overview of how sustainable fish guides are designed for use. http://overfishing.org/pages/guide_to_good_fish.php

3. Find two or three fish guides, either online or in stores that sell fish and seafood. Take notes on what you think about as you read the sources you selected. For example, what do you observe in each one that makes it a useful and informative fish guide? Do you notice any inconsistencies or confusing parts? How is the fish guide just right, or how could it be improved? Summarize your findings in a short written essay of about one page.

4. Bring your written summary and your notes to class to prepare for a group discussion. Having read the chapter and prepared the answers to the exercises, how has your perspective on the oceans changed?

7. THE POSSIBILITY OF GLOBAL GOVERNANCE

GASTON MESKENS

The following three exercises address different topics related to global sustainable development governance. They can be done each separately, but if your instructor assigns them all three together as a full lab assignment, it is advised to do them in the indicated order. Together they would enable you to learn more about how global governance happens today and about current visions on why and how it should and could be done differently. The preparation (task 1 of each exercise) can eventually be done as homework, but the focus of the exercises is on discussion in a small group.

Global Governance and the Millennium Development Goals

Use the websites of the United Nations (http://www.un.org/millenniumgoals/ and http://sustainabledevelopment.un.org/index.html) to find information on the Millennium Development Goals (MDGs). These goals meant to set into motion a number of specific practical policies to make the world a more just and sustainable place to live. The policies had to make sure that specific 'targets' were met by 2015.

1. Take time to study the meaning and content of each goal and try to understand why these MDGs were selected and how they interrelate. Then use the internet or any other source to get an idea of the various existing critiques about the fact that the MDGs were not realized.

2. Discuss your findings in a small group to learn from other students' views.

3. Then formulate and write down your own evaluation of the MDGs. Can you identify some reasons why they were not realized?

4. Come together as a larger group or full class again and discuss your own evaluation with the others.

Global Governance and the Sustainable Development Goals

Use the Sustainable Development website of the United Nations (http://sustainabledevelopment.un.org/index.html) to find information on the Sustainable Development Goals (SDGs). These goals are meant to replace the Millennium Development Goals from 2015 on and have the same ambition: to make the world a more just and sustainable place to live for all.

1. Take time to study the meaning and content of the goals and evaluate what are the main differences with the MDGs. What are the differences in focus? What are the differences in targets?

2. Discuss your findings in small groups to learn from other students' views.

3. Then formulate and write down your own evaluation of the SDGs and judge whether you find them realistic or not. Can they be achieved, and how? If you find them realistic, try to explain why. If you don't find them realistic, try to explain why not.

4. Come together as a larger group again and discuss your evaluation with the others.

Global governance and international politics

The negotiation processes on the MDGs and the SDGs facilitated by the United Nations are negotiations on a desirable future. Meanwhile the politics of global governance happen through strategic and diplomatic relations between the nation states of the world, and each of these states is primarily interested in acquiring (or maintaining) political sovereignty and a strong economic position. Today, there are two main visions on how global governance should be done: there are the voices who say that nation states will always think in their own interest first, which means that they will never be able to come to an ambitious but realistic agreement to eradicate poverty, tackle climate change or realise any of the other SDGs. For these people (often called 'cosmopolitans'), the only way out of this deadlock is the establishment of a strong international institute (such as an empowered United Nations) that can take over power from the nation states in order to motivate or even force them to work together. Other voices say that this is nothing more than a utopian dream and that we have to remain realistic and accept that nation states will never want to give up their sovereignty in the general interest. According to these people (often called 'realists'), eradicating poverty or tackling climate change will have to be done by each nation state alone on a voluntary basis. In addition, some of them also think that economic relations between the nation states in a global market can help those nation states with less capacity to fulfil their national goals in this respect.

1. Take time to think for yourself which of the two views you share and why. Write down a short consideration on why you think you are a cosmopolitan or a realist. Explain your view and your motivations.

   In order to find inspiration, or arguments to support your view, you can use the internet or any other source to learn more about strategic economic nation state networks (examples of these are the G8 and the G20) and international economic agreements (such as the Transatlantic Trade and Investment Partnership (TTIP)). In addition, it is essential to study international environmental agreements such as the United Nations Framework Convention on Climate Change (UNFCCC) or the Convention on Biological Diversity (CBD). In order to put your own view to the test, try to identify and evaluate how these networks and agreements can be critiqued.

2. Gather in small group and discuss each person's position. Is the group divided or is there a majority sharing one specific view? Summarize each person's position here.

3. For the views represented in your small group, what are the major components that group members identified? Summarize these major components in writing as a statement from your group.

4. From here, try to develop a joint written statement on how to tackle climate change on a global level. Whether you identify as a cosmopolitan, a realist or something else, what do you think should be done, and how can global governance be leveraged to make it happen?

Global Governance and the Environment

Choose one of the SDGS from the active UN goal discussion page which you can find under (https://sustainabledevelopment.un.org).

Some of the goals, such as those for food security, water supply, biodiversity, oceans, and sustainable terrestrial management are particularly suitable.

1. Read to learn about the scope of the goal. What topics and concerns are included or excluded from this goal?

2. Who are the major stakeholders in putting the goal together?

3. Who (or what) are the beneficiaries of this goal?

4. What targets have been identified for this goal? Evaluate the importance of each target.

5. What are the major challenges for successfully meeting this goal?

6. Write a briefing or report that discusses the findings of your research on this one goal.

SECTION 3
PARADIGM SHIFTS FOR SUSTAINABILITY

8. EDUCATION FOR SUSTAINABILITY

ELISE BOWDITCH

The following questions address different topics in the education chapter. Education is in some ways a more challenging topic because it is both abstract and practical in its application. We care about how to educate for sustainability, and what an education for sustainability is. The questions below are intended to get you thinking about how you have been educated and how the educational system you are currently in pays attention to sustainability. Since the chapter emphasizes the importance of simultaneous reforms in education worldwide, the questions ask you to connect with individuals and institutions that are working towards change for education and social sustainability. This is what we mean by a paradigm shift.

Your instructor may assign them to you as individual questions to be turned in as homework or a full lab assignment.

Collecting Data and Working with Data Analysis

POLICIES FOR SUSTAINABILITY IN SCHOOLS

Research your college or university's sustainability policies. You can explore the institution's website and find additional materials through newspapers, press releases, and other supporting materials.

1. Next, walk around the campus and look for examples of where the school does (or does not) adhere to its own policy. It is a good idea to talk with staff members involved in daily operations such as cafeteria workers, custodians, security guards, or athletic field staff to get a sense of the greater campus, not just a classroom building. It can be especially valuable to spend time understanding the heating and cooling operations and food handling at your school.

2. Take extensive notes on your observations and organize your notes for discussion in class.

Here are some questions to prepare you for your walk and get you started on your analysis.

1. Has your school signed any of the national university sustainability declarations?

2. Are they a member of any collegiate sustainability associations?

3. What types of food are offered on campus? Where does it come from? How much is wasted?

4. How much trash gets generated (daily, weekly, monthly) by dormitories, and where does it go?

5. Is there a recycling program? How closely do people follow it?

6. How does the school plan for and execute its energy needs?

7. What could the school improve regarding sustainability?

8. How well does the school live up to its own sustainability policies or statements?

Assignment

Write a report on your school's sustainability policy, the administration's efforts to support sustainability, and discuss how well your campus meets sustainability criteria in education.

This assignment can be written as an analysis and a critical evaluation based on your observations and notes. The assignment can be completed as an individual paper or as a group project.

Finding Sustainable Solutions:

PRACTICING SUSTAINABILITY IN SCHOOLS

Discussion

Think about your high school and prior experience as a student. Can you easily identify sustainable practices or attention placed on sustainability at the institution? Consider, for example, recycling, lighting, and water usage.

Questions for Discussion or a Homework Assignment

1. Did your school recycle? Was recycling effective?

2. Did any of your classes contain topics related to the environment? Which ones? How deep was the discussion?

3. Did your school support social criteria for sustainability such as student empowerment, student participation in creating curricula, school gardens, etc.?

4. What do you think are the barriers to bringing ecological or social sustainability efforts to a school? Do all high schools and colleges experience the same barriers? How could those barriers be overcome?

CONNECTING WITH THE GLOBAL SUSTAINABILITY NETWORK

It is often said that there is strength in numbers. While it is important for people and individual educational institutions to do their parts, we are always part of larger communities and networks. Normalizing sustainability as the way we do things involves education at every level of society, and much of that happens outside of schools. Many people want to live an environmentally friendly or completely sustainable, little- or zero-impact lifestyle. They find like-minded people through intentional communities such as ecovillages.

Assignment

Explore the Global EcoVillage website (gen.ecovillage.org/), and choose a community to examine more deeply. Some of them have their own websites, and many have contact information for new applications and questions at large.

Contact the community you would like to learn more about by phone or e-mail. Explain that you are a student and that you would appreciate their cooperation in helping you complete an assignment for class.

Investigate at least three topics from the following list:

1. Accepting new members and maintaining membership in the community
2. Children, child care, and schooling
3. Residential buildings
4. Nonresidential buildings
5. Waste management and recycling
6. Gardens and landscaping
7. Farming and food production
8. The individual and the community
9. Costs of maintaining the ecovillage
10. History of the ecovillage
11. Future plans for the ecovillage
12. The legal status of the ecovillage within the region or country

EXPANDING SUSTAINABILITY THROUGH AND BEYOND EDUCATION

In this chapter, you read about sustainable practices that are based on the teacher-student model of learning. This form of learning varies across countries, but typically occurs in a particular setting such as an elementary school and places learners in groups based on age.

How, then, can people be educated or trained in sustainability ideas if they are adults who finished school some years ago and never had access to sustainability ideas before?

This is the purpose of awareness campaigns or outreach to a specified community.

Assignment

Create your own awareness campaign for a sustainability-related topic about which you are passionate. There are two parts to the assignment. The first part is the rationale for why your campaign is needed. The second is to write an effective sustainability-awareness message for a specific target group you have identified in your rationale. You will need to consider which media to use and how to use them.

Begin by identifying the issue you want to focus on.

1. Next, make a list of at least ten ways to connect with adult individuals who are not in formal education settings.

2. Create a table with four columns. List your list of media outreach options in the first column. In the second column, identify the potential scope for each method. Estimate the cost of using or implementing the method in the third column. Finally, in the fourth column, identify the limits of the method.

   E.g.: Twitter is one media outreach method listed in column one.
   Second column: Global reach is possible, but hard to define and control.
   Third column: Cost is free to send, but needs subscribers to the feed.
   Fourth column: Limitations include time available to build a subscription base and language if English is used, but the target audience is Latino immigrants.

3. Identify which methods are most suitable for your issue and target audience, and write your awareness campaign.

9. ECONOMIC SCHOOLS AND DIFFERENT PATHS TO DEVELOPMENT

FRANKLIN C. GRAHAM IV

The exercises for this chapter ask you to reflect on the lives of real people as they are affected by global economics, and on the arguments different economic schools make regarding economic growth, social justice and environmental stewardship. Statistics can summarize a lot about a country while the economic (or geographic) reality they represent shape the way people make decisions.

Working with Maps and Data Visualization

SCHOOLS OF ECONOMIC THOUGHT

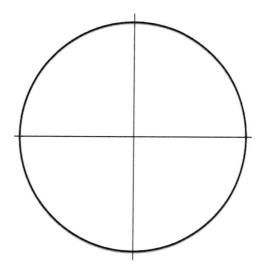

1. In the figure above, write in concepts and associated thinkers for each of the four major schools of economic thought talked about in this chapter.

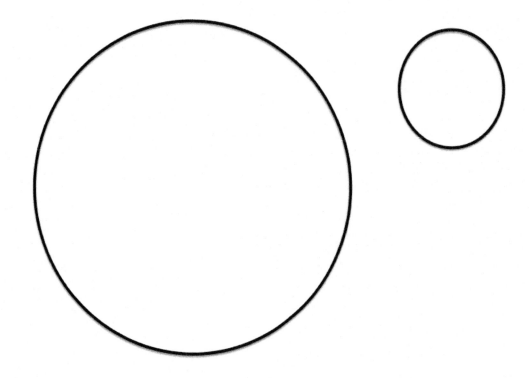

2. In the figure above, use the large circle to create a pie chart of the four economic systems. A pie chart is representational, so make larger slices for the economic systems in wider practice, and proportionally smaller slices for less practiced and marginal economic practices (those mentioned only briefly in the chapter). What goal or goals of sustainability (economic growth, social justice, environment) are pursued by the larger economic systems?
3. Use the smaller circle to identify economic practices that are more theoretical, or which critique the current economic systems in practice. What goal or goals of sustainability (economic growth, social justice, environment) are lacking in the global economic system and used in the critique of dominant schools of thought?

Working with Data Analysis

You read about Kuma and his decision to leave Ethiopia for the UAE. Compare and contrast the statistics below on Ethiopia and the United Arab Emirates. What statistics show evidence of Kuma's decisions regarding migration to the UAE? Are there any statistics that could indicate why Kuma plans to return to Ethiopia in the future? Write a short essay, using the statistics below as evidence that discusses your ideas as to why Kuma chose immigration to the UAE for employment, and also his reasons for planning to return to his home country. You may use additional sources to support a longer essay.

National Average	Ethiopia	United Arab Emirates (UAE)
Population*	96,633,458	5,628,805
Land Area (in square km)*	1,104,300	83,600
Climate*	Tropical Monsoon	Desert
Gross Domestic Product (Real Growth Rate)*	7 %	4 %
Per Capita Income*	1,300 USD	29,900 USD
Unemployment Rate**	17.5%	4.32% Overall (10.4% Nationals; 2.8% Non-Nationals)

Sources: * All statistics are from https://www.cia.gov/library/publications/the-world-factbook/

** Statistics for Ethiopia are from https://www.cia.gov/library/publications/the-world-factbook/

Statistics for the United Arab Emirates come from the UAE Ministry of Planning, 2014 (Abu Dhabi: UAE Government) retrieved on 7 October 2014 at: http://www.uaestatistics.gov.ae:81/ar/download.php?id=215

Economics, Politics and Society

1. What image of Neo-liberal economics does the U.S. media portray? Does it show favoritism for the Neo-liberal school of thought, is it critical of them, or well-balanced? Give a concrete example in your answer.

2. Name a U.S. law that reflects the influence of the economic reformist school of thought. Is the law representative of the Keynesian school, or does it draw in aspects from other schools? Justify your choice.

3. Given U.S. History, Communist candidates are not viable contenders for public office, but in other countries this is not the case. Pick a country where a communist political party or Marxist was or is active and explain briefly the contributions the party or individual made or makes in enacting Marxian development.

4. Radicals and anarchists are often associated with activism at organized protests at the meetings for the World Trade Organization and other large economic forums, and it is true, they come, but are not necessarily uniform in their message and tactics. What tactics do they employ to create awareness of their platform? Do their methods work? Explain your answer.

5. What common goal do all economic schools of thought share and when do they part in opinion? Do two or three schools of thought share similarities, and what tools (media, legislation, political parties, protests) do they use to promote their goals? In your answer, outline similarities and differences.

Economic and Social Sustainability

Kuma Baqqalaa expressed disgust regarding his occupation: hauling away the debris of perfectly sound homes razed for the development of high-rise apartments. Although Kuma moved past this comment, talking about how his job provides the means for his plans and goals, the observation is profound as it touches upon the issue of sustainability. It takes a large amount of energy, labor, natural resources and capital to convert the older neighborhoods of Dubai to high density residences. But even if this construction meets Dubai's housing needs, how sustainable is it to provide water, electricity and other services to these neighborhoods? How long can Dubai continue to fund this growth on petrol dollars and what does the extraction, processing and use of their oil and natural gas mean for the global carbon balance?

Part 1 : Collecting data and organizing your findings scientifically

For this assignment, identify the challenges the United Arab Emirates has in diversifying their economy and the impact of Emirati oil and natural gas on the global environment. In their effort to move away from natural resource extraction, is the UAE (1) seeking to create economic growth, (2) developing social justice within their laws, (3) proper stewardship of their natural environment or, (4) a balance of all three? What contributions does the oil coming from UAE make to emissions of greenhouse gases? What impacts does fossil fuel consumption have on weather, biodiversity, and people's vulnerability to natural hazards? Does the cost of gasoline (in various markets) reflect these costs?

Environmental Sustainability

Organize your findings from the assignment into a one-to-two page summary. Bring your notes and summary to class for a discussion. Putting aside the issues of economic growth and social justice temporarily, how do your findings regarding impact on the environment, both for the UAE and globally, compare with others in the class? How does the extraction of oil impact the immediate local ecology of the UAE and Persian Gulf?

Part 2: Discussing data and drawing conclusions

How does UAE oil and natural gas collaboratively burned with other fossil fuels impact the environment and communities in the Himalayas, the Maldives Islands, the eastern seaboard of the United States, the Sahel in Africa, the glaciers in the Alps and Andean Mountains, and the monsoonal patterns in South Asia? How do these climatic changes affect the local communities at the aforementioned regions economically and socially? What connections can you draw between climatic change and many of the conflicts over resources today?

10. WASTE MANAGEMENT: RETHINKING GARBAGE IN A THROWAWAY WORLD

DARRIN MAGEE

Working with Facts and Scientific Sources

EVERYDAY CONSUMPTION AND WASTE PRODUCTION

1. Examine your purchasing habits as a consumer over the past two years. Can you identify any of your purchases as at least partly motivated by a manufacturer's use of planned obsolescence as a business model? Explain the extent to which you felt you had a choice when making those purchases.

2. Examine the warning in Figure 10.3 of the main textbook from an environmental education center on a U.S. river. Do you think the sign effectively communicates some of the principles you have learned in this chapter? What does the sign suggest as preferred behaviors with respect to trash in and near waterways? Make a list of these suggestions, and add any you think would work even better.

WASTE MANAGEMENT TECHNOLOGY

1. Explain the "perverse ecology" of landfills and wastewater treatment plants.

2. Why is burn-barrel incineration of garbage more environmentally hazardous than industrial incineration?

3. Do you agree or disagree that electricity produced by WTE facilities should be classified as clean, renewable energy? Explain your reasoning. Why or why not?

Collecting Data and Working with Data Analysis

MANAGING TODAY'S WASTE STREAM

1. Use the Internet to find the location of up to ten landfills in your state. Websites you may wish to consult include the U.S. EPA and your state's own department of environmental conservation (or department of environmental protection). Which websites were most useful to you? Map your results on paper or by using one of the common mapping software tools. How close is the nearest landfill to your current home or school location? Does the location of your state's landfills make sense to you? Explain your answer.

2. Using the Internet and Google Earth or similar mapping program, map the distance from one major city to the three nearest landfills. What can you say about the geography of those landfill locations? Consider things such as topography, transportation routes, environmental considerations, and the economic or residential activities of nearby communities.

LEGACY CONTAMINATION AND CLEANUP

Assignments

1. Locate a closed landfill or "dump" near your home or school. Using the web, find out as much as you can about that landfill. When was it opened? When did it close? Was it lined? Why was it closed? What kinds of trash were buried there? Next, go to your local library and scan the local newspapers from that time period to see what you can find. What were residents and officials in your city or town saying about the landfill? Can you detect any common concerns such as environmental contamination or loss of jobs?

2. Many closed landfills in the United States are now on the SuperFund National Priorities List, a list of contaminated industrial sites for which federal SuperFund monies have been allocated for cleanup and remediation. Locate that list, and then use your ZIP code (or that of a friend or relative) to determine where the nearest NPL landfill is located. What are the key concerns and contaminants at that site? Where is it in the remediation process? Since this is all public information, what other information did you find interesting?

Finding Sustainable solutions:

PREPARING FOR TOMORROW'S WASTE

1. Visit the EPA's environmental justice page at http://www.epa.gov/environmentaljustice/. Look for resources that link to explore environmental justice outcomes near where you live. Then, choose a place where you expect environmental justice policies and outcomes to be better (or worse) than where you live. Use the research tools on the EPA site to see if your hunch was correct. How did the two places differ? What could be the reasons for these differences?

2. Use the EPA website or your state government's environmental agency website to find data about waste disposal. Try to locate historical data showing how disposal rates have changed by year. You may need to enlist the help of a reference librarian, because these data are not easy to find. Next, find historical data on per capita GDP for your state, or, more likely, for the entire country. Then, use a spreadsheet program to produce two line graphs, one showing waste generation (y axis) over time (x axis), the other showing per capita GDP (y axis) over time (x axis). What observations can you make about the shape of the two graphs? Finally, use the Web to learn about the Kuznets curve, and apply what you learn to your two graphs. Does waste generation in the United States follow the pattern a Kuznets curve might predict, or not?

3. Locate an area in any state where a landfill or incinerator project has been proposed for development or expansion. Next, examine local sources of public comment—newspapers, blogs, local government websites—to look for evidence of debate surrounding the proposal. Briefly characterize the key points of the debate and the groups espousing those points. Will citizens vote on the proposal? What is the approval process? Are you in favor of the new landfill or incinerator? What is your justification?

11. ECOLOGICAL LANDSCAPE PRACTICES: A SUSTAINABLE MODEL FOR NORTH AMERICA

MICHAEL WILSON

This chapter investigates ecological sustainability by looking at standard American methods for cultivating nonagricultural landscapes such as residential and commercial properties. The chapter reviews science-based changes in landscaping theory and practice. The questions below practice understanding and implementing the fundamental plant selection changes discussed in this chapter. The questions can be answered as individual assignments or as one cumulative project. Each piece can also be completed as a short paper that includes appendixes. All of the questions require additional research.

The sources for research may include state and local plant societies, botanical gardens and arboretums, land grant universities (cooperative extension service), the Environmental Protection Agency, the Department of Forestry, U.S. Department of Agriculture or USDA Plant Database, and Natural Resource Conservation Service (NRCS).

Collecting Data and Working with Data Analysis

PRACTICING ECOLOGICAL LANDSCAPE AWARENESS

1. Research to find the invasive species list for your state. Read any accompanying information, and evaluate the severity of the problem.

    a. Next, visit several local nurseries and garden centers with a copy of the list, and see if they sell any invasive plants or cultivars and varieties of invasive species.
    b. Develop a list of the most popular plants for sale that are also on the invasive species list.
    c. Find native alternative plants (perhaps with the assistance of nursery staff) that can serve the same purpose in the landscape.

*Note: New York State is developing a prohibited plant list to address invasive species and is allowing the nursery trade in New York a one-year grace period to sell off existing stock in listed invasive species.

2. Research to find the native plant species list for your area, county, or state.

    a. Prepare a copy for your class materials, citing sources as appropriate.
    b. Go to several local nurseries, and develop a list of the most popular native plants that are available at these nurseries. Cultivars and varieties should be included.
    c. Divide the list into trees, shrubs, and perennials, with notations on light and soil requirements.
    d. Next, investigate the landscape possibilities utilizing a layered, three-dimensional garden design that includes trees, shrubs, and perennials. Look for Internet sources that can recommend additional native plants appropriate for your climate zone that were not available at the local nurseries. Develop a list of plants that could be added to your landscape possibilities.

Working with Facts and Scientific Sources

PREPARING AN ECOLOGICAL LANDSCAPE FROM THE GROUND UP

1. Research and locate the storm water requirements of your local government, county, and state. Compare them with the requirements of the Water Quality Act of 1987 (Clean Water Act), and note the areas where compliance is strongest. Evaluate the data you collected.

    a. Find recommendations for rain gardens and suggested plants for your county or state. Check with local nurseries, and develop a list of the recommended rain garden plants that are available. Next, consult good-quality Internet sources, and create a list of the plants that were not available locally.

    b. Note any other methods of storm water mitigation that may have been suggested through your research like the use of pervious services, car washing and runoff, BMP, etc. Make a note of any overlap or gaps with the first batch of data you collected.

2. Obtain the soil survey for your county and find your location on the survey maps. Soil surveys are published by the NRCS or Soil Conservation Service. Some are available online; printed copies are available upon request. Copies are usually available in county libraries. The location you choose could be where you are living or the school campus. Find the soil types that are present at your locale. The soil types are noted in initials on the map. For example, on the soil survey map for Rockland County, New York, the initials HaA stand for Haven loam 0 to 3 percent slopes. Once the soils are located and the name is found, make notes of the general description of the soil. Soil surveys also contain a series of appendixes explaining the possibilities of the soils located in the survey, with charts for engineering, recommended agriculture for plants and livestock, development possibilities, and typical native trees growing on those soils.

   a. Write up the landscape possibilities for your location. Check for plant availability with local nurseries and Internet suppliers.

** Take this exercise a step further by having your soil tested for pH and nutrient analysis. Record the recommendations you receive, which may include a conventional approach. Find the organic equivalent to the conventional recommendations. Most land grant universities offer soil testing.

Finding Sustainable Solutions:

DESIGNING AN ECOLOGICAL LANDSCAPE

Create and draw your own ecological landscape on a real or imagined piece of property. Both options use the sketch map at the end of this chapter.

1. Draw a conceptual plan for an ecological landscape. You can enlarge the map provided with a photocopier, or redraw it. The sketch provides a house and a driveway. You can add the rest of the components to the yard. You may include any additional outbuildings like a shed, garage, or patio area. On the plan, mark off the types of gardens you would like in the conceptual plan like perennial, rain garden, or shrub border. Note how you will treat the lawn area by the type of seed or the alternative to the lawn. Note the percentage of lawn versus the size of the given lot. A conceptual plan offers a generalized plan and creates a vision for the property.

2. Draw a detailed design of an ecological landscape by noting on the sketch map the specific species of plants that will be used and their locations on the building lot. As you are plotting the specific species, consider the eventual mature size of the plants you are choosing, along with their characteristics. For example, you may not want to put an oak tree or another heavy fruit- or nut-bearing tree next to the driveway because of damage that may occur due to nut drop. A detailed design offers a more itemized, possibly step-by-step document on the specific and ideal plantings for the property.

Full-Term Assignment Option

Use all of the above information as a research background and the conceptual or detailed landscape plan to write a term paper or project. Along with your design, you will hand in a written component that will explain your design and your reasoning for selecting the plants you did. You can also include planting and maintenance recommendations such as herbivore protection, potential invasive plant threat, and invasive plant removal. Your report should also include several appendixes. One is a site evaluation. Another is a spreadsheet of the plants chosen that includes the number of plants, supplier, cost, wildlife benefits, and ornamental features. This plant listing may be broken down into two or three appendixes.

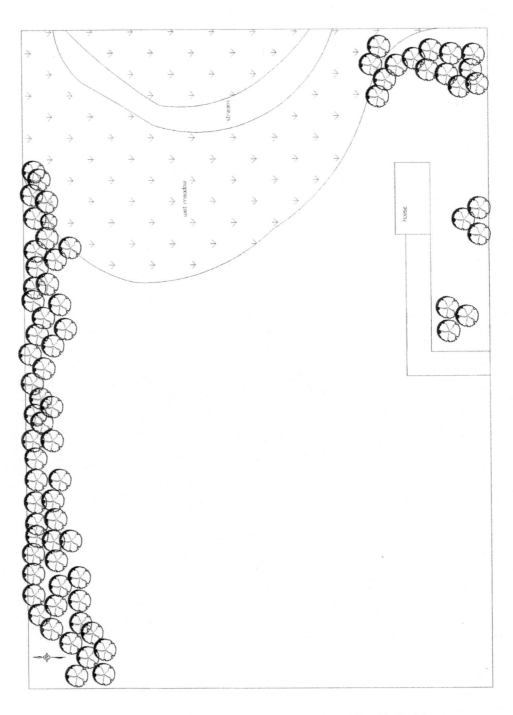

CPSIA information can be obtained at www.ICGtesting.com
Printed in the USA
LVOW02s1819070915

453176LV00002B/2/P